The INSIDE GUIDE

EARTH'S EXTREME PLACES

Ocean Trenches

By Judy Thorpe

Cavendish Square

Published in 2026 by Cavendish Square Publishing, LLC
2544 Clinton Street Buffalo, NY 14224

First Edition

Library of Congress Cataloging-in-Publication Data

Names: Thorpe, Judy, author.
Title: Ocean trenches / Judy Thorpe.
Description: Buffalo, NY : Cavendish Square Publishing, [2025] | Series: The inside guide: earth's extreme places | Includes index.
Identifiers: LCCN 2024057552 | ISBN 9781502674449 (library binding) | ISBN 9781502674432 (paperback) | ISBN 9781502674456 (ebook)
Subjects: LCSH: Submarine trenches–Juvenile literature. | Marine animals–Juvenile literature. | Deep sea biology–Juvenile literature. | Submarine topography–Juvenile literature.
Classification: LCC GC87.6 .T46 2025 | DDC 577.7/9–dc23/eng/20241230
LC record available at https://lccn.loc.gov/2024057552

Editor: Jennifer Lombardo
Copyeditor: Nicole Horning
Designer: Deanna Lepovich

The photographs in this book are used by permission and through the courtesy of: Cover Damsea/Shutterstock.com; p. 4 DOERS/Shutterstock.com; p. 6 Peter Hermes Furian/Shutterstock.com; p. 8 LukaSkywalker/Shutterstock.com; p. 9 DG-Studio/Shutterstock.com; p. 10 VectorMine/Shutterstock.com; p. 12 (left) Psychrolutes phrictus 1/Wikimedia Commons; p. 12 (right) Psychrolutes phrictus/Wikimedia Commons; p. 14 NOAA/Alamy Stock Photo; p. 15 Doug Perrine/Alamy Stock Photo; p. 16 Tunatura/Shutterstock.com; p. 18 imging/Shutterstock.com; p. 20 Dudarev Mikhail/Shutterstock.com; p. 21 DawidAndMarcelina/Shutterstock.com; p. 22 loocmill/Shutterstock.com; p. 25 courtesy of the Library of Congress; p. 26 Chronicle/Alamy Stock Photo; p. 27 (top) Operation 2021/Alamy Stock Photo; p. 27 (bottom) Titanic wreck bow/Wikimedia Commons; p. 28 (top) Ajm_a_l66/Shutterstock.com; p. 28 (bottom) Andrea Izzotti/Shutterstock.com; p. 29 (top) Dimitrios Karamitros/Shutterstock.com; p. 29 (bottom) zelvan/Shutterstock.com.

CPSIA compliance information: Batch #CSCSQ26: For further information contact Cavendish Square Publishing LLC at 1-877-980-4450.

Printed in the United States of America

CONTENTS

Chapter One: 5
Under the Sea

Chapter Two: 11
In the Trenches

Chapter Three: 17
A Dangerous Trip

Chapter Four: 23
Exploration and Research

Think About It! 28

Glossary 30

Find Out More 31

Index 32

Ocean trenches—also called deep-sea trenches—can be miles wide, so it is impossible to get a photo of the entire thing. This picture gives an idea of what a trench looks like.

UNDER THE SEA

Chapter One

Ocean trenches are deep **depressions** in the ocean floor. They are typically long and relatively narrow. However, even a "narrow" trench can be many miles wide. For example, the Mariana Trench—the most famous ocean trench—is about 43 miles (69 km) wide and more than 1,580 miles (2,540 km) long. The bottom of the trench, called the Challenger Deep, is more than 35,800 feet (10,900 m) below sea level. Ocean trenches are the lowest spots on the surface of our planet.

Trenches Around the World

Scientists know of at least 20 ocean trenches. However, because people have not yet explored the entire ocean, the exact number of trenches is unknown. After the Mariana Trench, the three deepest known trenches in the world are:

> **Fast Fact**
>
> The Mariana Trench is the most famous because it contains the deepest spot in the world. If Mount Everest were placed in the Challenger Deep, the mountain's peak would be about 7,000 feet (2,133 m) below sea level.

- Tonga: 35,702 feet (10,882 m) deep; located in the South Pacific Ocean

This map shows the location of the Ring of Fire and several ocean trenches.

Fast Fact

The word "trench" was first used to describe these areas of the ocean in 1923, when a professor of oceanography—the study of oceans—used it in his book, *An Introduction to Oceanography.*

- Philippine: 34,578 feet (10,539 m) deep at the Galathea Deep, the trench's lowest point; located in the western Pacific Ocean
- Kuril-Kamchatka: 34,587 feet (10,542 m) deep; located in the northwest Pacific Ocean

Most of the world's deepest trenches are located within the Ring of Fire. This is a horseshoe-shaped belt in the Pacific Ocean that is known for its earthquakes and volcanoes.

Plate Tectonics

The Ring of Fire is so geologically active because it is on top of the boundary of several different **tectonic plates**. Plate tectonics, or the movement of these plates, is responsible for the formation of geological features such as mountains, volcanoes, and ocean trenches.

Trenches form in a process known as "subduction." When two plates meet under the ocean, the one that is denser is subducted,

VOLCANOES AND ISLANDS

Subduction can also create volcanoes and island chains in and around the ocean. When one plate subducts under another, it pushes molten rock called magma to the surface, and the pressure of the magma creates a volcanic dome in the Earth's crust. Multiple volcanoes can form island chains.

Some trenches create accretionary wedges as they form. These are masses of sediment that get scraped off the denser plate as it subducts. The accretionary wedges pile up on the bottom of trenches. If there are enough of them, they can stick together and rise above sea level to create islands. This is how the Caribbean island of Barbados was formed.

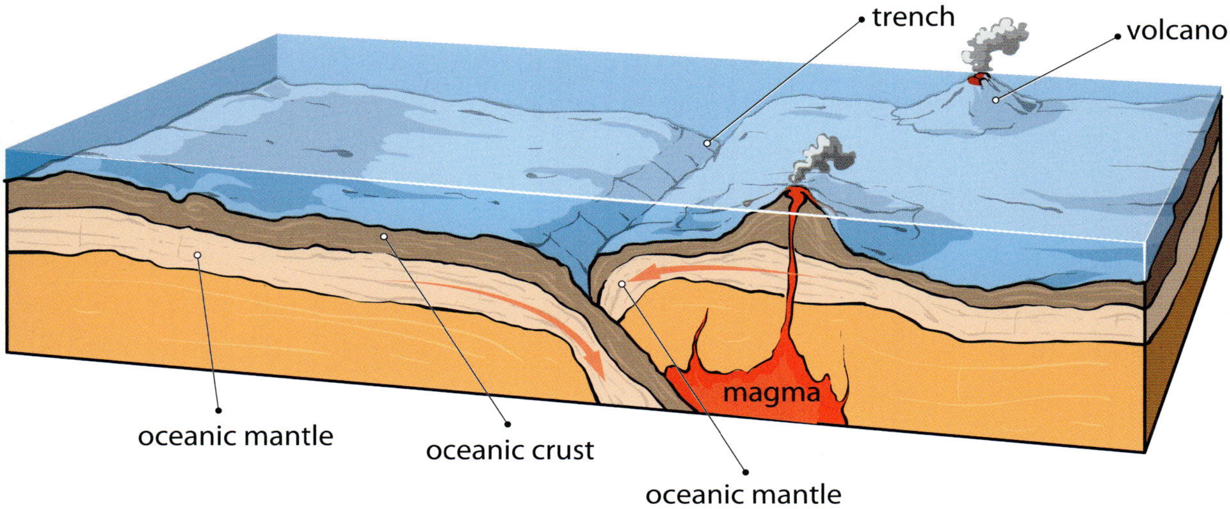

Fast Fact

Oceanic crust is denser than continental crust. If a continental plate and an oceanic plate meet, the oceanic plate will always subduct.

This picture shows how subduction creates ocean trenches and volcanoes.

or bent downward under the less dense plate. The place where subduction happens is called a subduction zone.

Nearly all oceanic subduction zones create a small hill before the trench. This hill is called the outer trench swell. It is the spot where subduction begins. The denser plate buckles or bulges before it is pushed under the plate that is more buoyant, or able to float. Scientists are not sure exactly what happens to a plate after it subducts. Some believe the heat of Earth's mantle melts it. Others think the subducted part of the plate sinks all the way into Earth's core before it melts.

Scientists have done experiments to try to figure out what happens during subduction, but it is difficult to know for sure. This

STRUCTURE OF THE EARTH

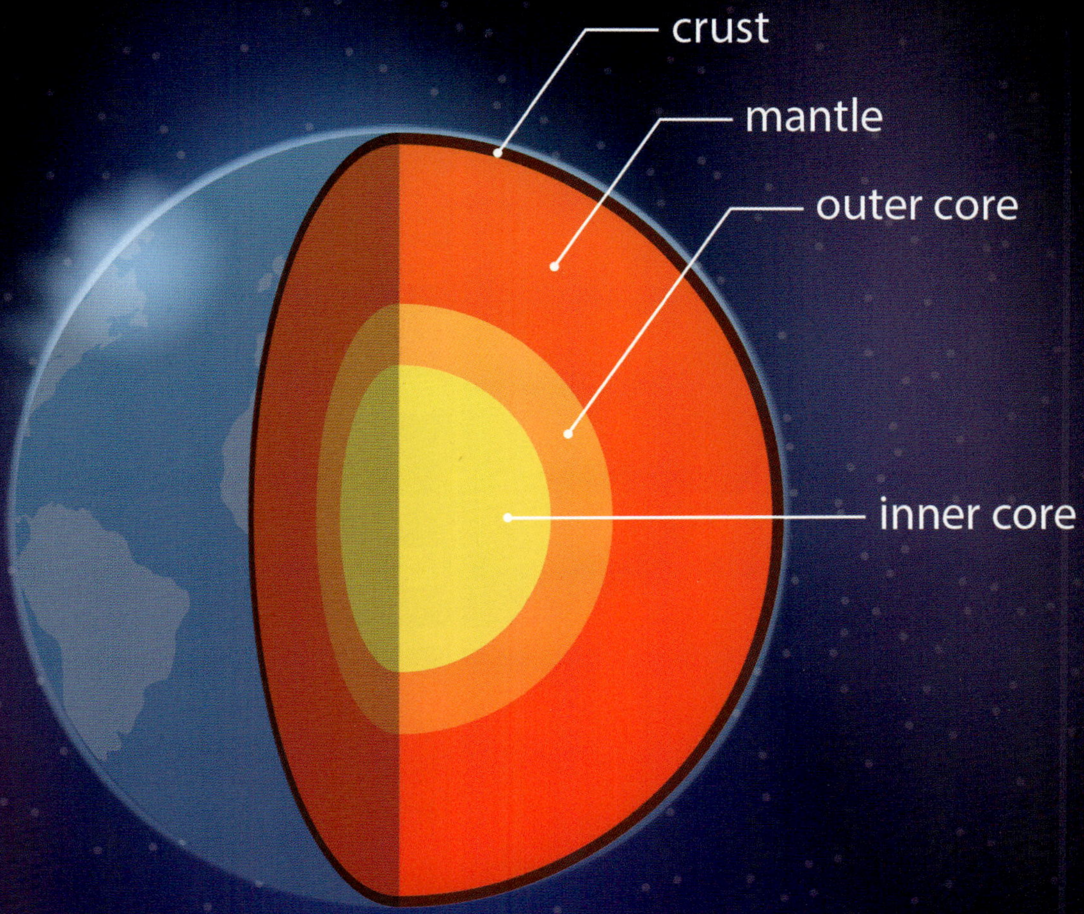

is because it is dangerous to travel to the bottom of the ocean and impossible to observe what is happening under the crust.

This picture shows Earth's layers. Tectonic plates form the crust.

OCEAN DEPTH

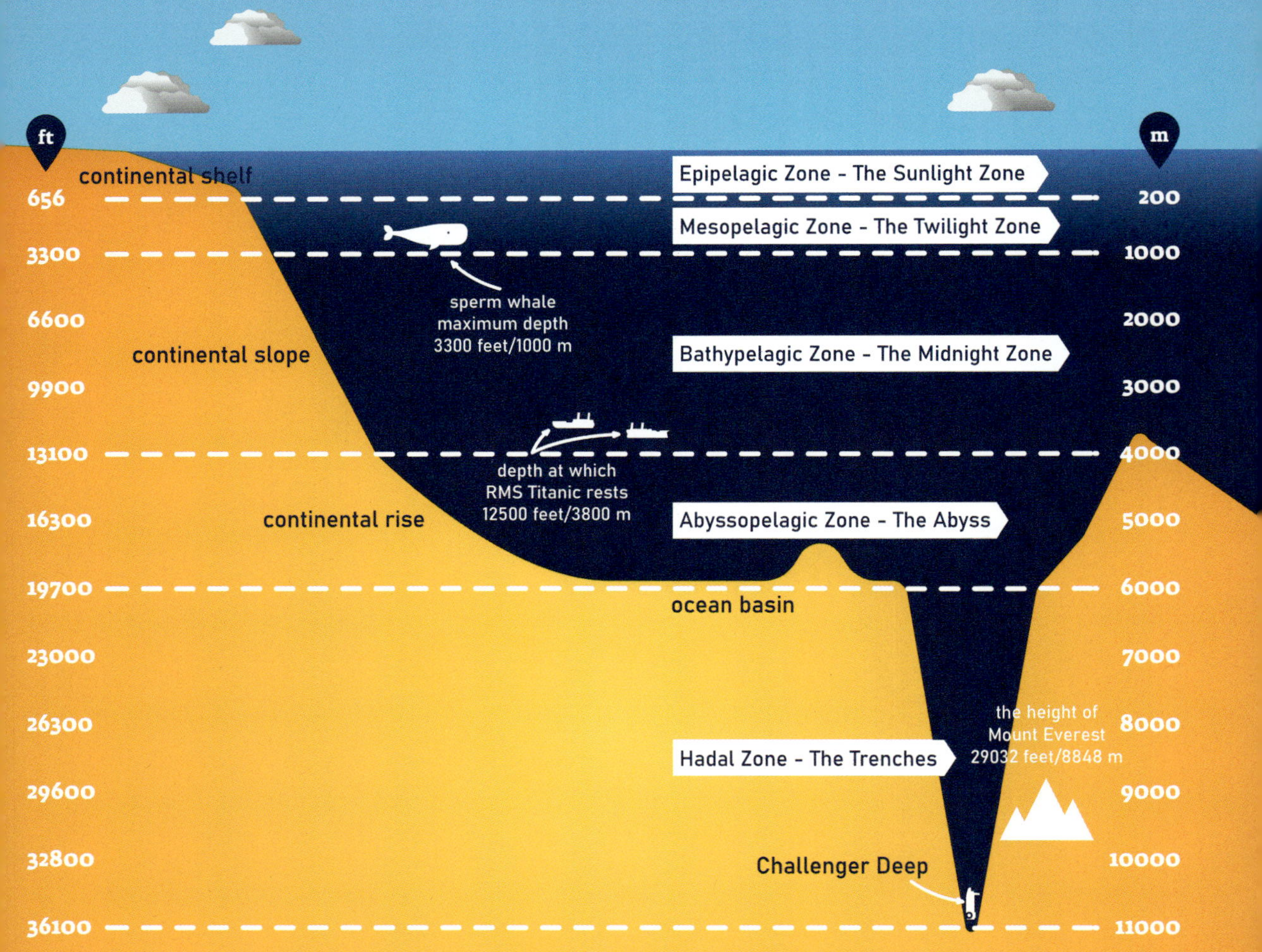

Scientists have separated the ocean into different layers, or zones, based on the amount of light each zone receives.

Chapter Two

IN THE TRENCHES

As the ocean gets deeper, it also gets darker and colder. Naturally, this makes trenches the darkest and coldest part. Additionally, the weight of all the water above creates **intense** pressure at the bottom of the deepest trenches. Very few animals and no plants at all can survive here.

Under Pressure

In the deepest trenches or parts of trenches, such as the Challenger Deep and the Galathea Deep, the water is just above freezing, and it is pitch-black. The pressure would squeeze all the air out of an animal that has lungs, such as a whale. For this reason, some creatures that live in deep trenches are **gelatinous**. They have adapted to the pressure, and it actually holds their soft bodies together. If they were brought to the surface, their bodies would collapse. Since people don't fish at those depths, this rarely

Fast Fact

The pressure at the bottom of the Mariana Trench would be deadly for a human to experience. A person standing in the Challenger Deep would feel about as much pressure as if 48 airplanes were stacked on top of them.

Blobfish live up to 4,000 feet (1,220 m) below the ocean surface. When a blobfish is underwater (*left*), it looks like a normal fish. When it is brought to the surface, the lack of pressure causes it to collapse (*right*).

Fast Fact

In 2022, scientists tried to bring a type of deep-sea jellyfish called a comb jelly to the surface so they could study it. However, they found that without the high pressure it was used to, the jelly's body completely disintegrated.

happens. However, the **phenomenon** has been observed often in animals that live in relatively shallower water, such as the blobfish.

Trench Life

Most animals that live in trenches look strange to us because they're so different than most animals that live closer to the surface.

One example is the goblin shark. It is sometimes called a "living fossil" because it has

BIOLUMINESCENCE

Bioluminescence is a characteristic of many animals that live in the deep sea. It is a result of a chemical reaction that happens inside cells. Some animals have all the cells that emit, or give off, light in a special organ called a photophore. This creates more intense spots of light. Animals without a photophore have light-emitting cells throughout their body. This gives them a softer, general glow.

The chemicals that create bioluminescence are called luciferin and luciferase. Some animals make these chemicals in their body. Some get them by eating other bioluminescent animals. Others have a **symbiotic** relationship with bioluminescent bacteria. The bacteria live in the animal's body, which gives them food and shelter. In return, they glow when the animal needs them to.

been around for more than 125 million years. Goblin sharks swim slowly. This helps them save energy—an important survival **tactic** in a place where food is hard to find. To help them catch faster-moving prey while still swimming slowly, their jaws can extend up to 3 inches (7.6 cm) out of their head.

Another animal that lives in ocean trenches is the vampire squid. This creature gets its name from its appearance, especially the webbing that connects its arms, which

Fast Fact

The vampire squid has the largest eyes proportionate to its body of any known animal in the world.

The dumbo octopus gets its name from its fins, which look like the ears of the Disney character Dumbo the elephant. It is the deepest-living octopus.

looks a bit like a vampire's cape. The vampire squid is not actually a squid, nor is it an octopus, but it has **characteristics** of both animals. All three, along with cuttlefish and nautiluses, are a type of animal known as a cephalopod.

Like the goblin shark, the vampire squid needs to conserve its energy. It is the only living cephalopod that scavenges, or eats parts of dead animals, instead of catching live prey. For this reason, it doesn't need to move very fast. Scientists believe its huge eyes help it see scraps of food as they float through the water.

The female anglerfish uses a bioluminescent light attached to its head to lure prey close enough to catch.

Scuba diving is an underwater activity in which people put on equipment so they can swim underwater for prolonged periods of time. The word "scuba" stands for "self-contained underwater breathing apparatus," and it describes the air tank divers wear.

Chapter Three

A DANGEROUS TRIP

The ocean is a dangerous place, but that has never stopped humans from exploring it as best they can. The following activities allow humans to experience deeper parts of the ocean, and with advances in technology, they may someday be able to happen in deep ocean trenches.

Diving

Diving is a popular way for people to explore ocean depths. The deeper a person goes underwater, the higher the pressure they feel. The human body is made up mostly of water, and water does not compress other water. This means that even at the bottom of the ocean, a person's body will not be crushed the way a soda can or Styrofoam cup would be. However, the gases in our body and the organs that contain them are compressed underwater. They decompress when the person returns to the surface.

Fast Fact

Water is about 1,300 times heavier than air. This is why the pressure of the water at the bottom of the ocean is so much greater than the pressure of the air at sea level.

THE BENDS

Divers need to be careful when they return to the surface from high-pressure depths. Doing so too quickly can cause decompression sickness—commonly known as "the bends" because people experiencing it bend over in pain. The air in a diver's tank contains nitrogen in addition to oxygen. Under high pressure, the diver's body uses the oxygen, and the nitrogen is absorbed into the diver's bloodstream.

As the pressure lessens, the body begins to clear the nitrogen from the blood. If the diver returns to the surface slowly, their body can do this a little at a time without much problem. However, if they return to the surface quickly, the nitrogen can form bubbles in the blood and muscles. These fizz like soda, block blood flow, and cause severe pain. The bends does not often kill people, but the pain it causes can last for days afterward, especially in the joints.

One popular kind of diving is called scuba diving. A person wears a wetsuit, mask, goggles, and a steel or aluminum tank full of pressurized air. Because they carry air with them, they can stay underwater for much longer

Some divers attach a decorated Styrofoam cup to their tank or suit. They keep the crushed cup as a souvenir.

than a person would normally be able to.

Fast Fact

The deeper a diver goes, the shorter the amount of time they can stay underwater before they start feeling sick. This has to do with the pressure, not the amount of air in their tank.

Scuba diving is relatively safe for someone with the right training, but it is not without risks. The biggest risk is drowning. If a diver panics while underwater, becomes confused and can't find the surface, or falls unconscious due to an unrelated medical emergency, they can drown. This is why people rarely dive without a buddy.

Another risk is a medical condition called an arterial air embolism, where an air bubble blocks blood flow in an **artery**. This can happen if a diver holds their breath instead of breathing normally through their mask, especially while ascending.

Another kind of diving is called free diving. In this activity, divers do not wear tanks. This means people who do free diving need to learn how to hold their breath for several minutes. Free diving takes away the risks of scuba diving that come from problems with the air tank. However, the risk of drowning still exists.

Fast Fact

A much more dangerous version, or type, of free diving is called no-limits free diving. Divers use weights to drop to dangerous depths and use gas-filled balloons or vests to shoot quickly to the surface. The record for no-limits free diving is 831 feet (253.2 m).

Some free divers use a rope attached to a buoy at the surface to help them dive deeper and ascend more slowly and safely.

Other Underwater Activities

Diving is the activity that involves going the deepest underwater. There are also a number of other underwater activities. These include:

- Underwater orienteering: Orienteering is an activity in which people use a map and compass to find their way in the woods without following a trail. In the underwater version, or type, people wear scuba gear and use a compass to follow an underwater route that has been set up in advance.
- Underwater wrestling: Two people try to be the first to remove a ribbon from their opponent's ankle underwater.

Coasteering is an activity that involves exploring a coast. While it does not take place fully underwater, part of coasteering that many people enjoy is jumping off high cliffs into the ocean.

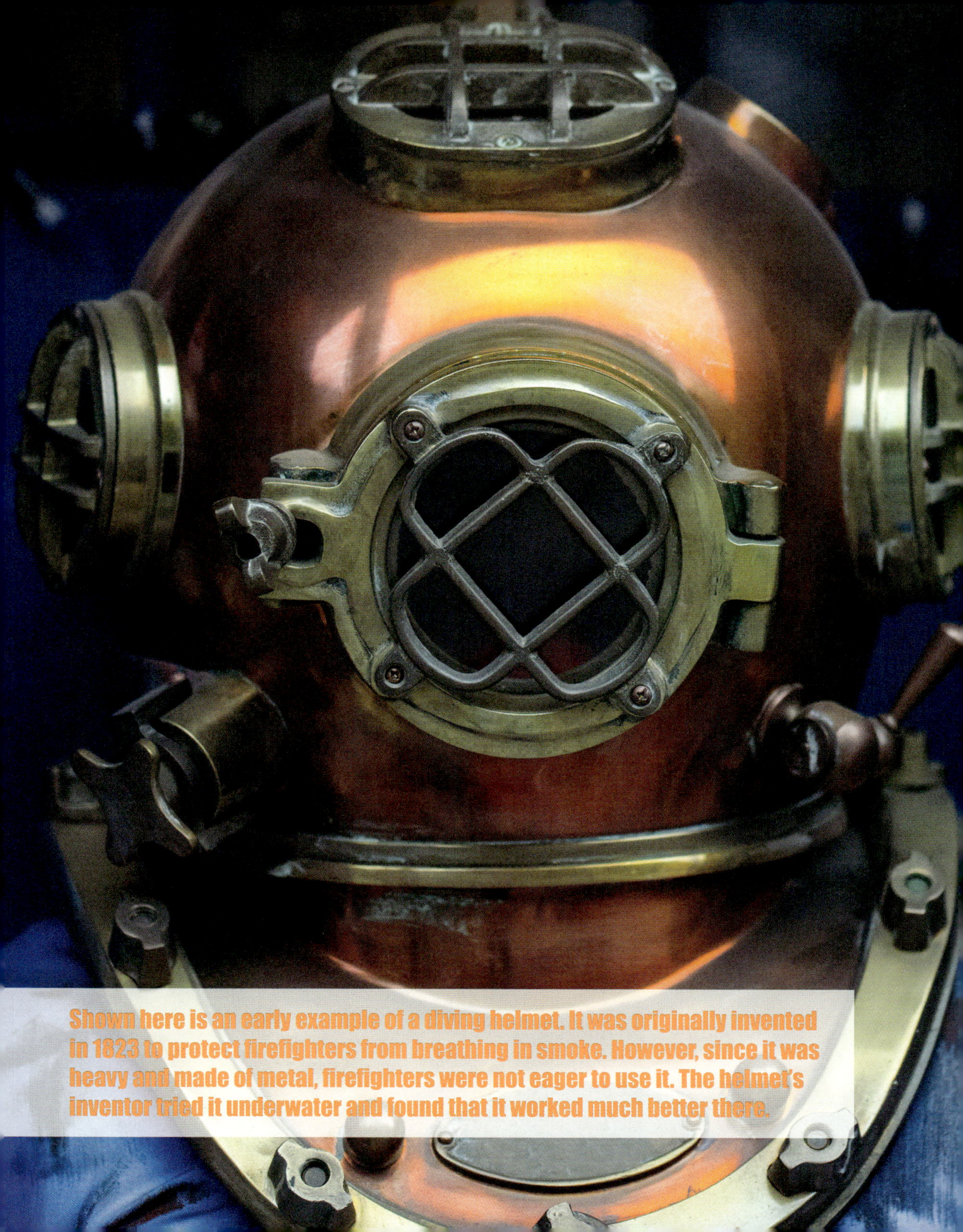

Shown here is an early example of a diving helmet. It was originally invented in 1823 to protect firefighters from breathing in smoke. However, since it was heavy and made of metal, firefighters were not eager to use it. The helmet's inventor tried it underwater and found that it worked much better there.

EXPLORATION AND RESEARCH

Chapter Four

People have been exploring the oceans since about 5000 BCE. However, because of the limitations of the human body, they could not dive very deep. Even today, this is a problem that takes a lot of work to overcome. Special suits and ships help, but even with these advancements, between 80 and 95 percent of the world's oceans remain unexplored—even though they cover 71 percent of the planet.

Oceanography

Oceanography is the scientific study of the ocean and the life it contains. This branch of science was started in the 18th century by an Italian man named Luigi Ferdinando Marsili. He spent a lot of time making observations of things such as the ocean floor, the coastline, ocean currents, and the properties of seawater.

Fast Fact

Early scientists often got things wrong. In 1843, an oceanographer named Edward Forbes confidently declared that life could not exist below 0.34 mile (0.54 km).

OCEANGATE

Exploring the ocean is not something to be done without care. Five people found that out in June 2023, when they took a trip in a submarine made by a company called Oceangate. The company's CEO, Stockton Rush, had been warned by multiple people before the trip that the sub was unsafe. There were several problems with it: Its hull, or outer part, was shaped in such a way that the water pressure was not distributed evenly across the ship; the hull was made of carbon fiber, which got weaker with every dive; and its window was only safe above 4,265 feet (1,300 m), but the sub was regularly diving to 12,500 feet (3,810 m) so passengers could view the remains of the *Titanic*.

Rush ignored the warnings and proudly stated that he was changing the way deep-sea diving had always been done. When the ship failed to resurface after a trip, an investigation found that it had imploded, or collapsed inward, killing everyone inside—likely instantly. Rush was one of the five people onboard.

Other scientists soon followed with their own observations, and eventually governments began realizing the importance of funding ocean research. At the same time, science was advancing, and people were inventing better tools for ocean exploration.

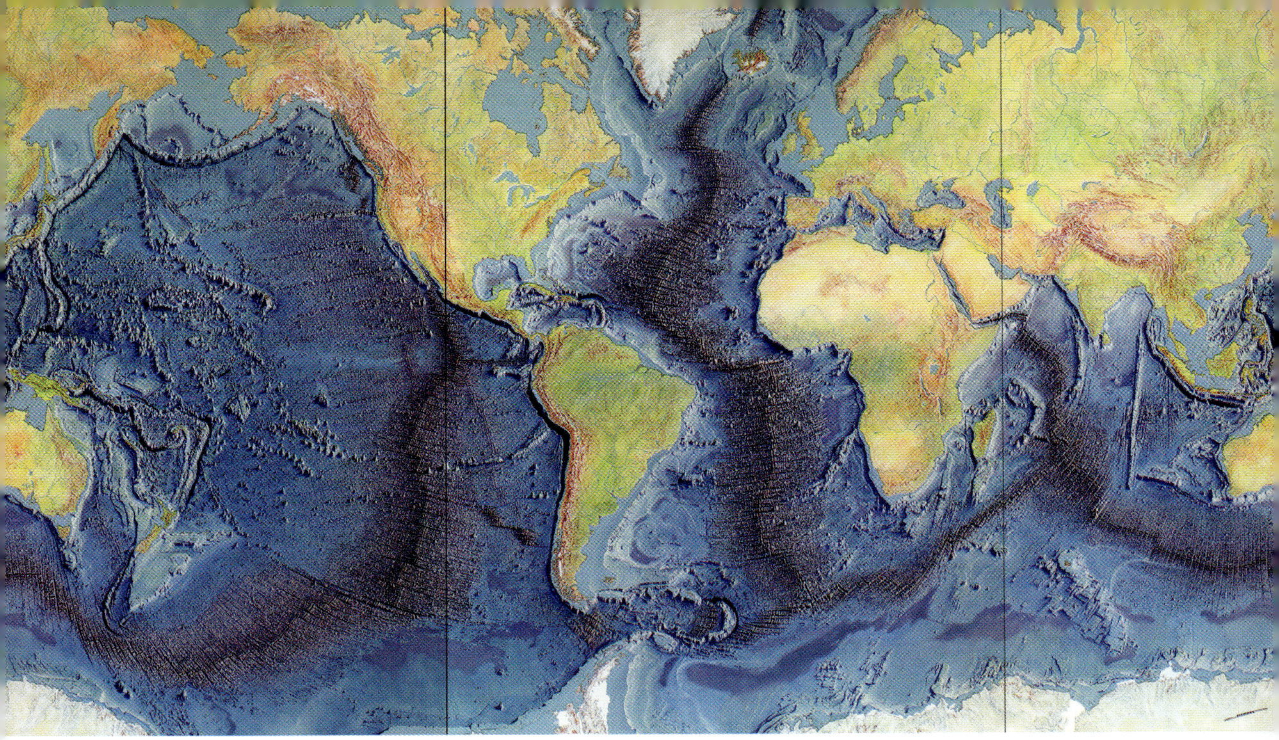

This map of the ocean floor was created in 1977.

Early Exploration

In 1930, two scientists named Charles William Beebe and Otis Barton created the earliest form of deep-sea ship, which they called a bathysphere. It was made of cast iron, had thick quartz windows, and was barely big enough for the two men to squeeze into it together.

The bathysphere was not pressurized the way modern submarines are, and it had no air supply, which limited the amount of time that could be spent in it. It also had no steering system. The bathysphere could be attached to the end of a long cable and dangled over the side of a ship. The deepest Beebe and Barton reached in the bathysphere was 3,028 feet (923 m).

The bathysphere was an important first step in the scientific exploration of the oceans, but it needed to be improved upon to get good results. For example, because the ship had no lights, it was

This picture shows Beebe climbing out of the bathysphere. Although he is often credited with its invention, it was actually Barton who designed the ship.

hard for Barton and Beebe to see the creatures swimming around them in detail. To get all the way to the bottom of the ocean, it also had to be much stronger, with the ability to be pressurized.

Hitting Bottom

After years of study and improvements by many different people and organizations, researchers were finally ready to attempt to reach the deepest part of the ocean floor. In January 1960, two men named Jacques Piccard and Don Walsh became the first humans ever to reach the deepest part of the Mariana Trench. This was never even attempted again until 2012, when director James Cameron made the trip. Cameron had also been to other parts of the ocean floor to visit the wreck of the *Titanic*.

Fast Fact

Beebe and Barton's trips were incredibly dangerous. If the bathysphere had burst under the pressure or if the cable holding it had snapped, both men would have died almost instantly.

In 2019, a company began offering regular dives to the Challenger Deep, and as of 2024, more than 20 people have made

Shown here is Don Walsh speaking at an event honoring the 60th anniversary of the *Trieste* dive. The ship itself hangs from the ceiling.

Fast Fact

The Challenger Deep gets its name from two British survey ships, *Challenger* and *Challenger II*. The ship that made the actual journey to the trench was called the *Trieste*. To withstand the pressure, the *Trieste*'s walls were 5 inches (12.7 cm) thick, and its windows were only 2 inches (5 cm) in diameter.

the trip. Some have gone for research, while others do it to set records or for fun. In 2020, Vanessa O'Brien, who has also successfully climbed Mount Everest, became the first woman in the world to reach both the highest and lowest points on the planet.

The *Titanic* is not in the Mariana Trench, but it is still quite far down at 12,500 feet (3,810 m).

THINK ABOUT IT!

1. What kind of experiments could scientists do to learn more about subduction?

2. Why can't plants survive in ocean trenches?

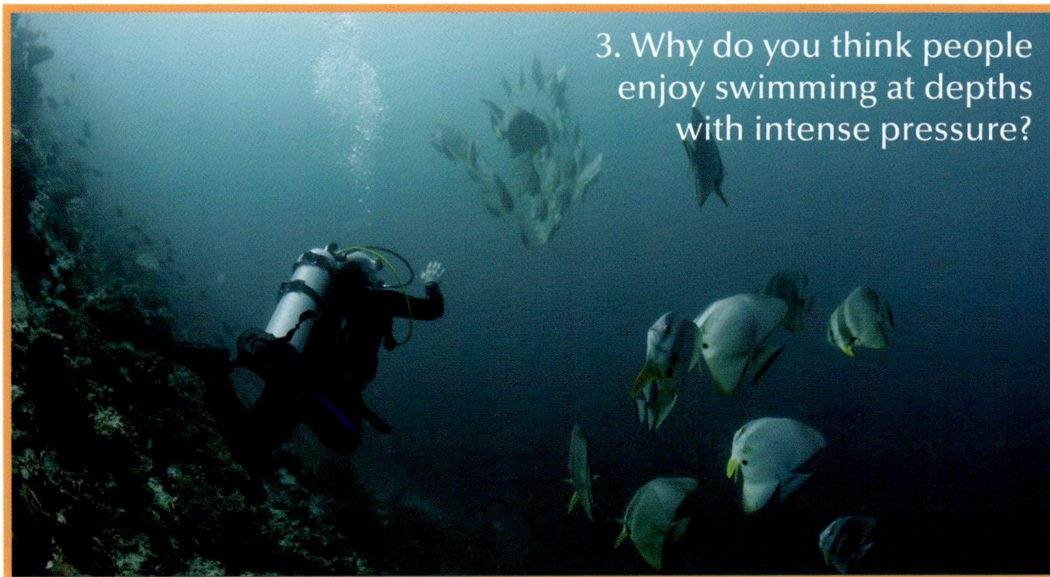

3. Why do you think people enjoy swimming at depths with intense pressure?

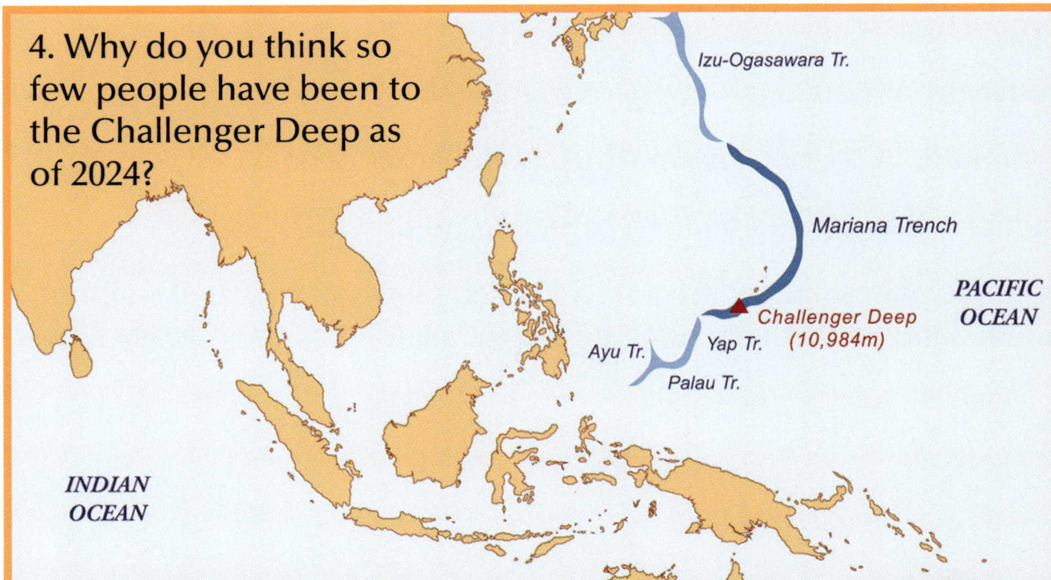

4. Why do you think so few people have been to the Challenger Deep as of 2024?

GLOSSARY

artery: A muscular tube that carries blood away from the heart to bring it to other parts of the body.

characteristics: Special qualities or appearances that make an individual or group different from others.

depression: A spot that is lower than the ground around it.

describe: To represent or give an account of in words.

disintegrate: To completely fall apart into tiny pieces.

gelatinous: Having the texture of jelly.

intense: Existing in an extreme degree.

phenomenon: A rare or important fact or event.

proportionate: Having a constant ratio.

souvenir: A reminder of a trip or event.

symbiotic: A relationship in which both parties benefit.

tactic: A planned action for a particular purpose.

tectonic plates: Huge, irregularly shaped pieces of rock that support all the land on Earth and create geologic features as they slowly move.

FIND OUT MORE

Books

Flannery, Tim F. *Deep Dive into Deep Sea: Exploring the Most Mysterious Levels of the Ocean*. New York, NY: Norton Young Readers, 2022.

Jackson, Tom. *Ocean Atlas: A Journey Across the Waves and Into the Deep*. Mission Viejo, CA: QEB Publishing, 2020.

Wasson, Karen. *Drop into the Ocean: A Tour of the World's Oceans and Seas*. Melbourne, Australia: Hardie Grant Explore, 2024.

Websites

BrainPOP: Ocean Floor
www.brainpop.com/topic/ocean-floor
Learn about trenches and more with Tim and Moby.

The Deep Sea
neal.fun/deep-sea
As visitors scroll to the bottom of this page, they learn about the life that exists at different ocean depths.

Woods Hole Oceanographic Institution: Ocean Trenches
www.whoi.edu/know-your-ocean/ocean-topics/how-the-ocean-works/seafloor-below/ocean-trenches/
Read more about how trenches form and why it's important for scientists to study them.

INDEX

B
bathysphere, 25–26
bends, the, 18
bioluminescence, 13

C
cephalopod, 14–15
Challenger Deep, 5, 11, 26–27

G
Galathea Deep, 6, 11
goblin shark, 12–13, 15

H
human body, 17, 23

I
island chains, 7

K
Kuril-Kamchatka, 6

M
Mariana Trench, 5, 11, 26

N
nitrogen, 18

O
O'Brien, Vanessa, 27
Oceangate, 24
oceanography, 6, 23
oxygen, 18

P
Philippine Trench, 6

R
Ring of Fire, 7

S
scuba diving, 17–19
seals, 11
subduction, 7–8

T
tectonic plates, 7
Titanic, 24, 26
Tonga Trench, 5

V
vampire squid, 13–15
volcanoes, 7

W
whales, 11